IMAGES
of America

HIGHWAY 61 THROUGH MINNESOTA

HIGHWAY 61'S MINNESOTA ROUTE. This map shows the approximate route of historic Highway 61 through Minnesota from the Canadian border at Grand Portage to the Wisconsin border in the southeastern corner of the state. The highway passes through many of eastern Minnesota's most prominent cities. (Courtesy of MNiCards.)

ON THE COVER: This photograph faces south on Highway 61 in downtown Pine City near Second Avenue South. The municipal liquor store and bar are on the right, and Ol' Smoothie Root Beer, a Pine City–brewed and bottled brand of root beer that started in 1953, is advertised on the billboard on the left. Highway 61 was the principal route from St. Paul to Duluth in the days before Interstate 35 ran through town. (Author's collection.)

IMAGES
of America

HIGHWAY 61 THROUGH MINNESOTA

Nathan Johnson

ISBN 978-1-4671-0693-1

Published by Arcadia Publishing
Charleston, South Carolina

Printed in the United States of America

Library of Congress Control Number: 2021937588

For all general information, please contact Arcadia Publishing:
Telephone 843-853-2070
Fax 843-853-0044
E-mail sales@arcadiapublishing.com
For customer service and orders:
Toll-Free 1-888-313-2665

Visit us on the Internet at www.arcadiapublishing.com

This book is dedicated to those who love their hometowns.
Mine, on "Old 61," is the greatest small town on earth.

Contents

Acknowledgments

There are many I would like to thank for the inspiration for this book. First, to the dedicated folks in east central Minnesota who are working diligently to reclaim and give a signal boost to Old Highway 61 through Carlton, Pine, and Chisago Counties, your work does not go unnoticed. Second, huge thanks to Cathy Wurzer for being the first to capture and share some of the many magnificent stories about the highway in her book *Tales of the Road*. The same goes for Jessica Lange, whom I thank for taking such lovely photographs and revealing them in her book *Highway 61*; it took me places vicariously that were reminiscent of my childhood and my many trips up and down the highway.

A big thank-you also goes out to the many others who helped enrich this project, including my hometown's Pine City Area History Association and the many other wonderful historical societies in cities and counties throughout Minnesota along 61.

Thanks to the wonderful places along the entire route that helped make my special 61 road trip in 2019 more amazing: Fair Trade Books in Red Wing, Best Western Plus in La Crescent, Hotel Julien in Dubuque, Main Street Bed & Breakfast in Hannibal, Cheshire Boutique Hotel in St. Louis, Hu. Hotel in Memphis, Shack Up Inn in Clarksdale, and Choctaw Hall in Natchez.

Thanks to my parents, and grandparents before them, and great grandparents before them, for choosing to live along such an amazing stretch of road that provided many good opportunities to me, taking me many places, and affording me many rich experiences. And most of all, praise and thanks to God, my maker, who granted me countless blessings and the knowledge, experience, and opportunity to take on this project.

Introduction

It feels so good to see the country, to give in to wanderlust and experience the vastness of America. Then again, Minnesota by itself has so much to offer. The feeling of being "up north" in Grand Marais is so vastly different than being in the bustling and historic capital city of St. Paul—though not even just from a size perspective. And both of those places are utterly different from the bluff country, in quaint river towns like Wabasha and Winona. They all feel like different worlds from one another.

That is why from the time I had a driver's license, I embarked on as many road trips as possible, especially regional trips throughout my home state, all of which started on Highway 61 near my childhood home. We lived just a couple of blocks away then, and I have since lived in six different places, all within blocks or directly on Highway 61.

As for road trips, there is something about the adventure of it: I love to go, see, do, taste, smell, connect, and stay awhile before embarking for the next destination. For example, a paper mill in Cloquet, Minnesota, permeates the air with a certain, distinct smell for miles. A truck stop in Rock Creek, Minnesota, is famous for its chicken dumpling soup, and an old brewery in Winona has been known for putting fried eggs on its hamburgers for generations. It is these unique, local quirks that identify a place that I love to experience. After all, "When in Rome . . ."

In elementary school, I treasured our class's time in the computer lab playing Oregon Trail. There was so much adventure that went along with traveling westward as a wagon leader from Independence, Missouri, to Oregon's Willamette Valley via a covered wagon, though I much prefer modern travel to that of 1848. For starters, there is less death now, and "hunting" for my meals consists of getting out my smartphone and typing in "best restaurants near me" rather than getting out a gun and some ammunition.

In high school, I discovered a new love: Sim City. I learned a plethora of useful information from that simulation game, such as supply and demand, budgeting, urban planning, and a ground-level (and underground-level) understanding of city utilities and services. It was then I learned the importance of connecting cities via highway. When a highway is built between cities, their citizens become more connected with the outside world. Better connections lead to more opportunities and more exposure to new ideas and other externalities, enriching the lives of those on either end of the highway.

Later, my time as a community development director and city planner in Pine City helped me relate what I had learned to real life. I came to understand that, without US Highway 61 running directly through downtown, the city would not have flourished as much as it has through the years. I worked with others to help exploit the old highway and bring attention to the amenities along it. After Interstate 35 bypassed the community, I worked with the American Association of State Highway and Transportation Officials and local and county officials to designate Old 61 as an official I-35 Business Loop through Pine City to encourage motorists to get back onto the original highway and re-explore all of what Pine City has to offer.

It was my fondness of seeing new cities and small towns that made me ripe for a good road trip. Many people prefer to just get there, hence the growth in air travel, but you miss so much not seeing all of the "in between." A gravestone has the beginning and end dates connected by a dash. The richness of one's life is in that dash, the "in between." A Linda Ellis poem reads, "For that dash represents all the time they spent alive on earth and now only those who loved them know what that little line is worth."

If you take the freeways all the time, you do not get an understanding of what the highways and byways are worth. I loved discovering alternate routes from the Rand McNally road atlases of yesteryear and, more recently, taking "Avoid Highway" routes on my navigation app and seeing all of the dots in between. There are lots of sayings about the journey rather than the destination. With that notion, my best friend Annie and I hit the road in the fall of 2019 with the aim to drive Highway 61 in its entirety and have a "Huckleberry Finn" adventure of our own—minus the racism. We drove all 1,400 miles of it, crossing eight states and largely following the Mighty Mississippi from one end to the other.

Soon after, the great pandemic hit. We did not know just how important it was to take our trip when we took it—until we looked back. We loved being able to freely interact with the friendly locals in Dubuque, dine in some of the best Italian restaurants in the Hill in St. Louis and the mouth-watering barbecue joints in the alleys of Memphis, and walk in close proximity to the crowds and street performers in the French Quarter of New Orleans. Much of what we did, we would have experienced differently, if at all, had COVID-19 struck earlier.

We would do the trip all over again. It was the trip of a lifetime. We came to realize the vastness of Highway 61 in its entirety and the prominence it has had in the telling of America's story. We loved so many of the wonderful cities and towns and sights along the way. Taking in cities such as St. Louis, Memphis, Clarksdale, and New Orleans on the "Blues Highway" portion of the route, we had a relatively easy drive. Even with all there was to do and see elsewhere, it made us especially proud of our Minnesota portion of Highway 61 and all it has to offer.

There is no denying Route 66 is still the classic road trip in the United States, or at least the one more Americans point to. That said, Highway 61 makes a superb alternative, and it spans the entire country, north-to-south, not just half, like its more famous east-west cousin. Highway 61 is America's north-south byway, and it undoubtedly grants motorists a one-of-a-kind, awesome experience.

One

The North Shore Drive

Grand Portage to Duluth on Minnesota 61

The stretch of Highway 61 between the Canadian border and Duluth is arguably one of the most picturesque portions of highway in the United States. Commissioned in 1926 and ready for use by 1929, it bestows views of the shoreline of Lake Superior in all of its glory. It was not paved until 1940. Marked now as Minnesota State Highway 61 (MN 61) or North Shore Scenic Drive, it beckons one to get behind the wheel and drive its 140-plus miles over and over again.

From the Canadian border, Grand Portage comes first. Along the way, motorists look out their windows to see the Sawtooth Mountains, North America's largest lake, prominent lighthouses, and many charming small towns. An added bonus to the scenery is the waterfalls, notable among them Gooseberry Falls and High Falls.

The route is also part of the Lake Superior Circle Tour running through Ontario, Michigan, Wisconsin, and this part of Minnesota.

This route was US Highway 61 until 1991, connecting Ontario, Canada with New Orleans, Louisiana, but everything north of Duluth became MN 61 after that portion was turned back to the state. Music legend and Duluth native Bob Dylan referred to this road in his notable sixth studio album, *Highway 61 Revisited*.

From Two Harbors to Duluth, MN 61 is a four-lane expressway established in the 1960s and later designated the Arthur V. Rohweder Memorial Highway. Rohweder dedicated his life toward saving others' lives. He served in many capacities in the realm of public safety and welfare in Minnesota and was nationally prominent in all areas of accident prevention work.

Alternatively, Scenic Highway 61, or County Road 61 (CR 61), is a scenic drive running virtually parallel with the expressway but more closely abutting Lake Superior.

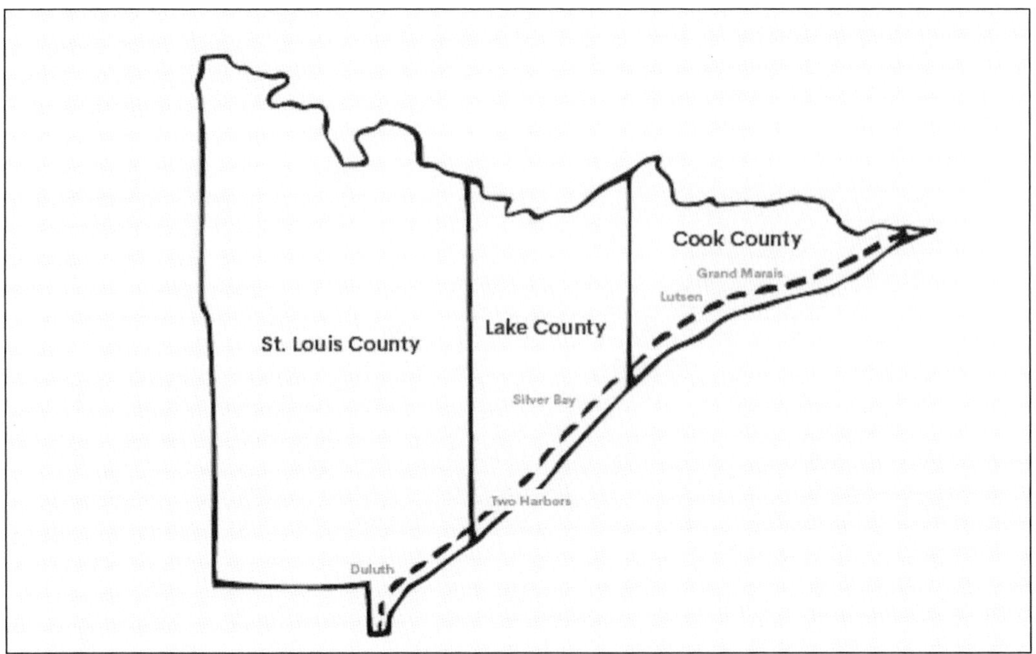

Grand Portage to Duluth. This map illustrates the route of Minnesota State Highway 61 through Cook, Lake, and St. Louis Counties, from the Canadian border at Grand Portage, 152 miles to Duluth. This part of northeastern Minnesota is very scenic, and the majority of this portion of Highway 61 offers views of Lake Superior, the largest of the Great Lakes. (Courtesy of MNiCards.)

Outlaw Bridge. The "Outlaw Bridge" was the northernmost point of US Highway 61. The original bridge crossing the Pigeon River from Minnesota to Ontario, it was constructed without federal approval by the rotary clubs of Duluth and Port Arthur (now Thunder Bay). It opened to traffic on August 18, 1917, creating the first road link into Port Arthur, and came down in 1930 after falling into disrepair. (Courtesy of Cook County Historical Society.)

Pigeon River. The Pigeon River serves as the Canada–United States border along Highway 61. In pre-industrial times, it was of great importance for transportation and fur trading. The river flows over 30 miles east out of the Boundary Waters Canoe Area Wilderness and drains into Lake Superior. The Pigeon is one of the larger rivers on Superior's north shore. High Falls, pictured here, is the 120-foot transitional waterfall defining the border. (Courtesy of Cook County Historical Society.)

Grand Portage. The Grand Portage is an 8.5-mile footpath that bypasses waterfalls and rapids on the final 20 miles of the Pigeon River before it flows into Lake Superior. It was a historical trade route of the French-Canadian voyageurs between their wintering grounds and their depots to the east. Today, a trail exists along Grand Portage in much the same location and condition as in historic times. (Courtesy of Tony Webster.)

GRAND PORTAGE NATIONAL MONUMENT. Grand Portage National Monument and Indian Reservation form a bridge between people, time, and culture. For over four centuries, Ojibwe families have tapped maples, fished, and harvested in this area. There was a rich exchange of fur and ideas between the Ojibwe people and the English and French fur traders. The North West Company had its headquarters here from 1778 to 1802. (Author's collection.)

HOVLAND DOCK. About 20 miles northeast of Grand Marais, on Chicago Bay in the present-day ghost town of Hovland, Minnesota, lies the last remaining vintage commercial dock on Lake Superior. It dates to the early 1900s, when the town was thriving. The dock's primary use was for loading and unloading passenger and cargo ships traveling between Canada and Duluth. It stretches out about 100 feet into Lake Superior. (Courtesy of Cook County Historical Society.)

NANIBOUJOU LODGE. Naniboujou Lodge is a resort built as a private club on the north shore of Lake Superior, 15 miles northeast of Grand Marais. Named after a Cree Indian character, its décor has influences of both Native American and Art Deco. Grandiose plans succumbed to the economic realities of the Great Depression; only a small clubhouse was built that retains its original design. It is listed in the National Register of Historic Places. (Courtesy of August Schwerdfeger.)

ST. FRANCIS XAVIER CHURCH. St. Francis Xavier Church is colloquially known as the "Chippewa City Church" because of the once-thriving Chippewa City, where it is located. Today, it stands as a reminder of that once vibrant place of more than 100 families in the 1880s and 1890s. George Morrison, a Native American artist best known for his brilliantly colored landscape paintings and wood collages, was born in the village. (Courtesy of Minnesota State Historic Preservation Office.)

Gunflint Trail. Once a primitive trail used mostly by travelers on foot or dogsled to move goods for trade to the port of Grand Marais, Gunflint Trail is now a 57-mile paved road that begins at Highway 61. A national scenic byway since 2009, the route winds through Superior National Forest at the edge of the Boundary Waters Canoe Area Wilderness. Today, it is a hub for camping, canoeing, hiking, and fishing. (Courtesy of Cook County Historical Society.)

Grand Marais. Grand Marais is the first significant town motorists arrive at on Highway 61 after leaving Canada. Originally a Native American village and a fur trading hub, it was incorporated in 1903 and remains a charming village that appeals to outdoor enthusiasts who appreciate having Lake Superior at the city's doorstep and Superior National Forest in its backyard. (Courtesy of Cook County Historical Society.)

Street View of Grand Marais. Grand Marais loosely translates from Ojibwe to English as "big marsh." For generations, locals have logged and fished the area. The county seat of—and sole municipality in—Cook County, Grand Marais has a wide variety of restaurants, shops, lodging, and attractions despite its small size. (Courtesy of Cook County Historical Society.)

Grand Marais Art Colony. Artists have lived and interacted with one another since 1947 at Minnesota's oldest art colony, the Grand Marais Art Colony. Started by acclaimed regional artist Birney Quick as a summer painting program, the colony grew to serve over 25,000 people in a variety of art mediums. It now also hosts over 200 classes a year and contains a shop filled with intriguing items by artists in residence. (Courtesy of Cook County Historical Society.)

Lutsen Resort. Lutsen, an unincorporated community with just under 200 inhabitants, is home to some of the north shore's largest and most iconic attractions, with several right along Highway 61. These include the historic lodge at Lutsen Resort, Lutsen Mountains ski area, and Superior National Golf Course. Lutsen, which is within Superior National Forest in Cook County, is a hub of recreation. It was the first ski resort in Minnesota and home to the first swimming pool in the world to be enclosed in a Mylar bubble. (Above, courtesy of Jereme Rauckman; below, courtesy of John Frid.)

Carlton Peak Lookout Tower, Tofte. Tofte is an unincorporated community in Superior National Forest in Cook County. Located about 27 miles southwest of Grand Marais on Highway 61, recreational opportunities abound in the vicinity. Carlton Peak Lookout Tower and the Gitchi-Gami State Trail are nearby. Named after a community in Norway of the same name, Tofte has been around since 1898. (Courtesy of Superior National Forest.)

Edgewater Inn and Olsen's Motel. The Edgewater Inn was a motel and store in Tofte. On the same property was Olsen's Motel for a time. Both were torn down, and Dennis Rysdahl built condos on the land in 1984. Rysdahl's development became a popular Lake Superior resort called Bluefin Bay. Ownership transitioned to Joe Swanson in 2020. (Both, courtesy of John Frid.)

SCHROEDER LUMBER COMPANY BUNKHOUSE. Schroeder is an unincorporated community in Cook County on the banks of the Cross River, 30 miles southwest of Grand Marais. Outdoor enthusiasts enjoy Temperance River State Park and the Superior Hiking Trail nearby. Schroeder was named after lumber baron John Schroeder. The nearby Schroeder Lumber Company bunkhouse, part of a former logging camp, is listed in the National Register of Historic Places. (Author's collection.)

SILVER BAY'S "ROCKY TACONITE." Silver Bay is a north shore town of 1,800. A roadside attraction just off Highway 61 serves as the town's mascot, Rocky Taconite. Wearing a miner's hardhat, holding a pick, and sitting atop a taconite boulder, he represents the industry that drew thousands to this community in the 1950s. Rocky was completed in 1962 and was moved here in 1964. (Courtesy of Mel Wilson.)

Beaver Bay. On a curve along Highway 61 sits Beaver Bay, the oldest settlement on the north shore of Lake Superior. It was established in 1856, two years before Minnesota, and tourism has long played a large role here. The community is home to the annual John Beargrease Dog Sled Race, named after the community's most famous resident, an Ojibwe mail carrier and fur trader who traveled long distances via wooden dogsled. (Courtesy of Lake County Historical Society)

Gooseberry River Bridge. This bridge along Highway 61 over the Gooseberry River was constructed in the 1920s to meet the needs of the emerging tourist economy along the north shore of Lake Superior. Three waterfalls called Gooseberry Falls are popular stopping points for tourists. The bridge was replaced in the 1990s with a similarly styled bridge. (Courtesy of Library of Congress.)

Silver Creek Cliff. North of Two Harbors, the portion of Highway 61 at Silver Creek Cliff was a treacherous portion of the roadway prior to the 1920s because of how close it went to the edge, near Lake Superior, as it went around the cliff. Many motorists would take a detour several miles inland. The roadway was improved in the 1920s and a tunnel, not finished until 1994, replaced the dangerous roadway. (Courtesy of University of Minnesota–Duluth Kathryn A. Martin Library.)

Gooseberry Falls. Gooseberry Falls, northeast of Two Harbors, is close to the community of Castle Danger. Three Norwegian anglers settled there in 1890, but it was occupied earlier by lumber companies. There is some debate about how it got its name: the anglers may have thought the cliffs along the shore resembled a castle, a boat named *Castle* may have been grounded there, or it could come from the 1831 Walter Scott novel *Castle Dangerous*. (Courtesy of Joe Haupt.)

Split Rock Lighthouse. Eight miles south of Silver Bay, Minnesota, sits Split Rock Lighthouse, which offers tours of it and the keeper's home by guides in 1920s clothing. Considered one of the most picturesque lighthouses in America, it was featured in the 2013 film *The Great Gatsby*. It was built in 1910 due to the great loss of ships during the Mataafa Storm of 1905, in which 29 ships were lost in Lake Superior. (Courtesy of Library of Congress.)

Betty's Pies. Betty's Café (later Betty's Pies) began in 1958 after replacing a fish shack Betty Lessard's father built along 61 north of Two Harbors. Betty made anglers donuts and coffee to start. Later came a popular strawberry pie from a newspaper recipe. Pies eventually made the cafe a landmark. Betty died in 2015 at age 90, but Betty's Pies became a staple for motorists on their way to cabins and resorts. (Above, courtesy of Amy Meredith; below, courtesy of John Frid.)

Two Harbors. This postcard was produced in the 1940s by Curt Teich and depicted several recognizable Two Harbors scenes. Included in the big letters were images of wildlife, the harbors, an ore dock, a downtown street scene, a lodging option of the day, and the Two Harbors Lighthouse, among others. (Courtesy of John Frid.)

Two Harbors Main Street. Just a few blocks off of the main highway, MN 61, is Main Street. Two Harbors was developed as a railroad and iron ore–shipping town. It became the first iron ore port in 1884, and was incorporated in 1888. Ojibwe people had hunted and fished the area long beforehand. (Courtesy of John Frid.)

Lou's Fish House. Trips through Two Harbors often include a stop at Lou's Fish House, once named by *Martha Stewart Living* the best smoked-fish house in America. The fish include trout from nearby Lake Superior, as well as whitefish and herring. The original owner, Lou Sjoberg, was also a pipefitting welder and a professional wrestler. (Courtesy of Chad David Capp.)

Two Harbors Lighthouse. Minnesota's oldest operating lighthouse is the Two Harbors Lighthouse, which overlooks Lake Superior's Agate Bay. Construction began in 1891 to provide safe passage into the harbor, a major shipping point for iron ore in the late 1800s. First lit on April 14, 1892, three keepers kept it lit around the clock. Besides its role as an active lighthouse, it now serves as a museum as well. (Courtesy of Pete Markham.)

Voyageur Motel. The Voyageur Motel has accommodated motorists traveling along Highway 61 for decades. For most of its existence, it drew visitors with a campy voyageur statue named "Pierre," who stood there from 1960 to 2011, when he was relocated to the nearby Earthwood Inn. Pierre had eyes with slits that glowed and moved, and a head that moved back and forth as he talked to guests. (Courtesy of Terry Johnson.)

House of Sweden. Walt Grant built a restaurant called the House of Sweden in 1946. It served cuisine from his native land. Later, in the 1960s, he added motel rooms. He sold the business to Tony and Marge Radosevich in the 1970s; they converted the property to the Earthwood Inn, which still operates as an inn, restaurant, and bar. (Courtesy of John Frid.)

North Shore Scenic Drive. The North Shore Scenic Drive was constructed between 1922 and 1923 and follows the original path of Highway 61 between Duluth and Two Harbors. It provides motorists a scenic alternative to the four-lane expressway completed further inland in 1967. The old road provides views of Lake Superior and passes through the small towns of Talmadge, French River, Palmers, Knife River, and Larsmont. (Courtesy of John Frid.)

Tom's Logging Camp. Tom's Logging Camp has been a quirky roadside attraction since 1956, providing insight into Minnesota's logging history. Displayed inside are many random objects, such as an electric chair, a Paul Bunyan cutout, and a room full of chainsaws. It even has funhouse mirrors inside a crooked building. (Courtesy of Bart Heird.)

Fish Fry Lodge. Fish Fry Lodge was on the shores of Lake Superior along Highway 61, about 10 miles north of downtown Duluth. It advertised 24 modern cabins housing from two to eight people, and "also twin beds," according to its signature postcard. Other amenities included hot and cold water and golf and riding facilities nearby. It was open each year from May 1 to November 1. (Courtesy of University of Minnesota–Duluth Kathryn A. Martin Library.)

Glensheen Mansion. The posh, 39-room Glensheen Mansion began as Chester and Clara Congdon's estate in 1908. Chester came to booming Duluth in 1892 and became a key figure in iron mining and banking, as well as a state legislator. He became the wealthiest Minnesotan. Prior to a notorious murder on the premises, the Congdons willed Glensheen to the University of Minnesota–Duluth. It has since become one of Minnesota's top historic tourist attractions. (Courtesy of University of Minnesota–Duluth Kathryn A. Martin Library.)

Duluth Armory. The Duluth Armory along 61 was constructed in 1915 at nearly five times the cost of other armories at the time. A training facility for the Minnesota National Guard and Naval Militia, it also put Duluth on the map as a cultural and entertainment hub. Harry Truman, Buddy Holly, Louis Armstrong, Johnny Cash, Bob Hope, Liberace, and more appeared there. It is now an arts and music center. (Courtesy of University of Minnesota–Duluth Kathryn A. Martin Library.)

Bob Dylan Way. Bob Dylan, born Robert Zimmerman, is a Duluth native. The singer-songwriter, who is known for such songs as "Blowin' in the Wind" and, notably, his sixth album, *Highway 61 Revisited*, was born in "the Zenith City" in 1941 and left a major mark on American history. In 2006, the city designated a 1.8-mile cultural pathway in honor of Dylan, commemorating his 65th birthday, winding through downtown on 61. (Courtesy of Ross Heister.)

Fitger's Brewing Company. Fitger's brewed beer from 1881 to 1972, but its brewery complex still sprawls along 720 feet of Lake Superior's shoreline and historic Highway 61. It is comprised of 10 buildings constructed between 1886 and 1911, which now serve as an indoor mall with shops, restaurants, nightclubs, a hotel, and a museum about Fitger's. An exhibit about Bob Dylan Way is also on display. (Courtesy of Sharon Mollerus.)

Aerial Lift Bridge. Duluth's Canal Park has a high concentration of tourist amenities such as restaurants, shops, and lodging. Tourists have primarily flocked to this part of the city for the Marine Museum and lake walk along the shores of Lake Superior, as well as the Aerial Lift Bridge. The bridge lowers and raises over 5,000 times per year. (Courtesy of Library of Congress.)

Enger Park and Tower. Overlooking Duluth since 1939, Enger Tower offers panoramic views of the city's skyline. A visible landmark from Highway 61, it sits 531 feet above Lake Superior. It is named after Bert Enger, who emigrated from Norway first to Pine City and then Duluth, where he became a successful furniture retailer. A majority of Enger's estate was left to the City of Duluth, including the parkland the five-story tower sits upon. (Courtesy of University of Minnesota–Duluth Kathryn A. Martin Library.)

Hotel Duluth, Greysolon Plaza. Hotel Duluth was built in 1925, and with its 14 stories and 500 rooms, it was considered one of the finest hotels in North America. In 1980, it was converted in part to apartments, while the ballroom and other spaces were made available for special occasion rentals. Notable guests have included Henry Fonda and Pres. John F. Kennedy. (Courtesy of Nick Benson.)

DULUTH UNION DEPOT. The Union Depot replaced Duluth's original depot and began operations in 1892, as the city was expanding and the need for a larger passenger station came about. At its peak, it handled 60 trains per day. In the 1970s, it became in part the Lake Superior Railroad Museum after being listed in the National Register of Historic Places. (Courtesy of University of Minnesota–Duluth Kathryn A. Martin Library.)

Two

TWIN PORTS TO TWIN CITIES

DULUTH TO WYOMING ON OLD HIGHWAY 61

The first section of Highway 61 to be completed was between the Twin Cities (Minneapolis–St. Paul) and Twin Ports (Duluth-Superior), which was paved by 1927. It was upgraded and modernized between 1946 and 1957.

The pre-interstate alignment of US Highway 61 from Wyoming to Duluth was bypassed between 1962 and 1974, when Interstate 35 became the primary route between the Twin Cities and Twin Ports.

Travel time through Carlton, Pine, and Chisago Counties (the pine logging region of east central Minnesota) was reduced with the development of the new highway. Some of the small towns like Pine City and Hinckley expanded toward the freeway; others, such as Rock Creek and Beroun, virtually vanished.

Hinckley business owners caused a stir during the turn-back process in 1962. When the rerouting of 61 to the freeway happened without consulting the local residents, a temporary alignment through town returned until the state developed a more formal process.

The pre-interstate alignment of 61 is nearly intact today except for at major curves. It is a two-lane highway, save for the gateway to the Duluth metro area near Scanlon, where it became a four-lane divided highway. Mostly, a string of county roads occupies the old highway for the remainder of the route.

The Old Highway 61 Coalition formed in recent years in the three counties to promote and preserve Old Highway 61 within Carlton, Pine and Chisago Counties. One of its projects has been to work with the counties to install "Old" Highway 61 signage along the route.

Many of the iconic stops and attractions along this stretch of highway have stood the test of time. Some have since vanished. For example, just four miles south of Pine City in Rock Creek, at the intersection of State Highway 70 and Highway 61, was the Pine Outdoor Drive-In. Owned and operated by the Gross family, it had one screen and a capacity of about 300 cars. The structure was demolished to make way for a logging company that was established in 1998.

In 2016, a stretch of 61—from First Avenue North in Pine City to Everready Road/County Road 55—was officially designated by Pine County as Deputy Sheriff Mike Morrow Memorial Highway to commemorate the 25th anniversary of the death of Morrow, killed in the line of duty in 1992.

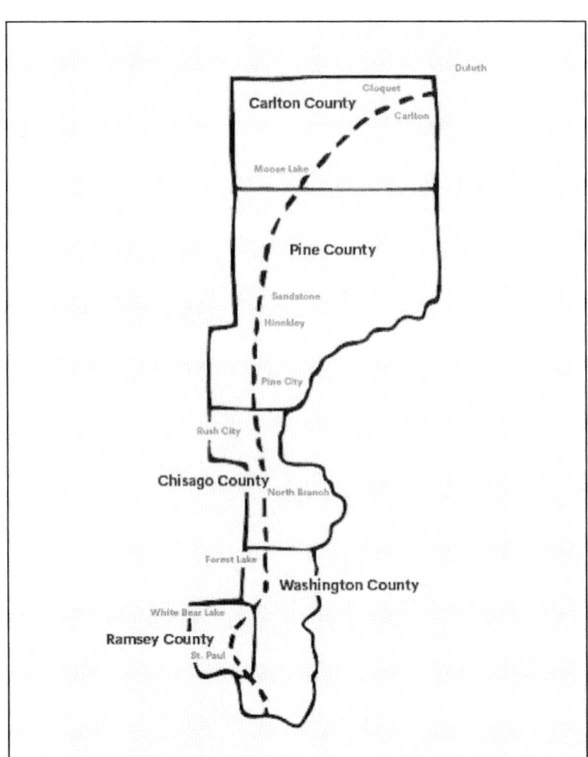

Twin Ports to Twin Cities. This map shows the approximate route of Highway 61 and its role as an important connection between the state's two largest metropolitan areas, Duluth-Superior and Minneapolis–St. Paul. Here, the highway passes through Carlton, Pine, Chisago, Washington, and Ramsey Counties in east central Minnesota, west of the St. Croix River. Cities along the way include suburbs such as Cloquet, White Bear Lake, and Forest Lake, and more rural small towns in between like Hinckley and Pine City. (Courtesy of MNiCards.)

Carlton Main Street. Highway 61 used to run right through the center of Carlton, but that route was decommissioned. Carlton was a railroad town on the Northern Pacific Railroad route from Duluth to St. Paul. Temporary shacks were erected for the railroad workers here at Northern Pacific Junction, later incorporated as Carlton in 1881, after state senator Rueben Carlton. (Courtesy of John Frid.)

TJ's Country Corner. Mahtowa is home to a real, old-style country store, TJ's Country Corner, famous for its bratwurst and potato sausage. "TJ" comes from Tom and Joanne Bislow, who opened it in 1977. Though Tom passed away in 2019, his plethora of activities and marketing ideas for Mahtowa and his store breathed life into the town. (Courtesy of TJ's Country Corner.)

Carlton County Fair. Barnum is 30 miles southwest of Duluth and five miles northeast of Moose Lake on Highway 61. It was named after George G. Barnum, a paymaster of the region's intra-city railroad. The city was incorporated on February 5, 1889. Since 1891, it has been the host of the annual Carlton County Fair. Since the fair's second year, it has offered horse racing, to the delight of people around the area, from Esko to Moose Lake. While many county fairs in Minnesota used to offer horse racing, the Carlton County Fair is the only one that carried on the tradition. (Courtesy of Carlton County Historical Society.)

Barnum Saloon. Ole and Emma Peterson operated the Barnum Saloon right next door to their home in downtown Barnum. In 1900, friends of theirs gathered at their home, on both levels, for a photograph while they were getting ready for a big dance to take place on the unpaved street. (Courtesy of Carlton County Historical Society.)

Lou's Rustic Diner. Lou's Rustic Diner, in Barnum, opened in 1979, serving meals for motorists traveling along Highway 61, and serving as a community gathering place ever since. In 2012, the diner was heavily flood-damaged, but the local community rallied to support and help Lou and Robin Paulson reopen it. This important piece of Barnum's identity remains today. (Courtesy of David Cooley.)

Moose Lake. Moose Lake was incorporated in 1889. Since 1969, it has held Agate Days, a weekend festival featuring rocks and an agate stampede in the center of town. The city is considered the agate capital of the world due to past glacial activity that formed an abundance of the rock. In fact, the world's largest agate, 108 pounds, is on display in a downtown bank window. (Above, courtesy of University of Minnesota–Duluth Kathryn A. Martin Library; below, courtesy of John Frid.)

Moose Lake Station. The Moose Lake Station was a Soo Line Railroad depot built in 1907, replacing a depot constructed there in 1873. It was one of the few structures to survive the Cloquet Fire of 1918. Today, it serves as a museum run by the Moose Lake Area Historical Society, and the Soo Line is now a recreational trail. (Courtesy of Ron Reiring.)

YMCA Camp Miller. YMCA Camp Miller was started in 1898 as the ninth YMCA camp in the United States. It began in Ely, and over the years it has been relocated, eventually to Sturgeon Lake just off Highway 61. The camp serves horseback riders, families, and youth, with a focus on boys and girls aged 7–16. It is the sixth-oldest remaining YMCA camp in the nation. (Author's collection.)

Peggy Sue's Café. The building that houses Peggy Sue's Café in Willow River has a varied history since it was built in 1895. It has been a post office, doctor's office, and a temporary schoolhouse after the town's first school burned in a fire. It first became a café in 1924, the Willow Café and Confectionary. In 1995, Peggy Sue and Al Villa took over her brother Van's business, and Peggy Sue's was born. (Courtesy of Martin Rainer.)

Willow River Rutabaga Plant. Rutabagas are a celebrated root vegetable in northern Pine County. Askov celebrates Rutabaga Days, and a number of storing and processing facilities for rutabagas existed until some were burned by fire. Willow River Rutabaga Warehouse and Processing Plant, built in 1935–1937, still stands along Highway 61 and is listed in the National Register of Historic Places as the state's only surviving agricultural facility dedicated to storing and processing rutabagas. (Courtesy of Minnesota State Historic Preservation Office.)

Banning Junction. Just off Highway 61 is a multi-purpose building called Banning Junction, which has had a number of owners and uses over the years, most often as a fueling station and supper club. Banning, now a ghost town, became Banning State Park in the 1960s, and Banning Junction served as a gateway to the park. (Courtesy of Brian Henderson.)

The 61 Motel. A classic roadside motel, 61 Motel has existed on the north end of Sandstone, on the west side of Highway 61, since 1955. While guests are primarily weary road trippers, the accommodations also serve friends and family of inmates at the nearby Sandstone Federal Correctional Institution. Actor and comedian Tim Allen spent two years and four months at Sandstone until he was paroled in 1981. In 1975, Scenic Sign Company of Pine City was commissioned to construct a neon sign for the motel. The iconic sign has since been fully restored yet hangs slightly crooked, with a bright-yellow arrow. (Both, courtesy of Jerry Huddleston.)

Midwest Country Music Theater. What came to be known as the Vogue Theater replaced Sandstone's opera house, which burned in a fire on the 300 block of Commercial Avenue. In 1997, the 300-seat venue became the home of Midwest Country Music Theater. In recent years, performances are regularly televised on RFD-TV. (Courtesy of Paul VanDerWerf.)

Sandstone Depot. The Great Northern Railroad passed through Sandstone, and downtown formed nearby. Originally called the Village of Fortuna in 1857, it served as the county seat for Buchanan County. Sandstone, just north of Fortuna, was platted in 1887. Three years later, the two villages merged and reincorporated as Sandstone. Railroad tycoon James J. Hill constructed many of the structures in Sandstone. (Courtesy of John Frid.)

Sandstone Quarries. Just off 61 in Sandstone, on the banks of the Kettle River, is Robinson Park. This is where extensive quarries of Sandstone were first worked in 1885, leading to settlement in the area. Since 2005, ice climbers descend on the park each winter for the Sandstone Ice Festival. (Author's collection.)

Hinckley Main Street. The Ojibwe people were the first to settle what is now Hinckley. In 1869, the first railroad came through Hinckley, and that began a major expansion of the railroad in Central Station, as the town was known from 1885 until it was incorporated as Hinckley in 1907. Main Street became a commerce hub, running perpendicular to Highway 61. (Courtesy of John Frid.)

TOBIE'S. Tobie's began in 1920 as a restaurant serving donuts and coffee on the corner of Old 61 and Main Street in Hinckley. In 1947, Tobie and Ann Lackner became proprietors and changed the name to Tobie's Eat Shop & Bus Stop. The restaurant moved to near the freeway in 1966 and became world famous for its baked goods, especially its king-sized caramel and cinnamon rolls. (Both, courtesy of Chris Hickle.)

Hinckley Fire Museum. The Hinckley Fire Museum on Highway 61 interprets the ravaging Great Hinckley Fire of 1894 and the subsequent rebuilding of the town. A quarter-million acres burned in just four hours. The official death toll was 418, not counting hundreds of Native Americans who lived in and around the town and others who were never found. The story was told around the world. (Above, courtesy of Andrew Munsch; below, courtesy of János McGhie.)

HINCKLEY CORN & CLOVER CARNIVAL. The Corn & Clover Carnival, Hinckley's crown jewel event, has taken place off and on since 1911. There were several years throughout the Great Depression the event did not take place. A well-known photographer, Ross Daniels, who had a studio in nearby Pine City, took this postcard photograph in 1920. (Courtesy of John Frid.)

WILLARD MUNGER STATE TRAIL. The Willard Munger State Trail starts in Hinckley just off Highway 61 and goes northeast to Duluth. From the trailhead to the Twin Ports, it is 63 miles in length, and ranks as the fifth-longest paved trail in the United States, reclaiming an old railroad that was used to save many lives during the Hinckley and Cloquet fires of the 19th century. Hikers, bicyclists, in-line skaters, and snowmobilers enjoy the trail named after a legislator devoted to trail development during his legislative career spanning from 1954 until he died in 1999. Munger holds the record as the longest-serving member of the Minnesota House of Representatives, at 42 years and seven months. (Above, courtesy of Steven; right, courtesy of the Minnesota Legislative Reference Library.)

FRANÇOIS THE VOYAGEUR. François began as a 42-foot-tall redwood log, which sculptor Dennis Roghair used to create America's largest redwood sculpture in the early 1990s for Pine City. It represents the voyageurs, French Canadians who transported goods to and from the nearby Snake River Fur Post, and has become a popular roadside attraction. (Courtesy of City of Pine City.)

SNAKE RIVER FUR POST. West of 61 in Pine City is the Snake River Fur Post, a reconstructed post that was originally established in the fall of 1804 by the North West Company's John Sayer and his crew of voyageurs. Artifacts turned up in the 1930s led to the discovery of the site. Today, motorists traveling the highway make this a stop for cultural education as well as recreation. (Author's collection.)

Petchel's and Nicoll's Café. The building that now houses Nicoll's Café began as a saloon in 1897, owned by brewer Theodore Buselmeier. During Prohibition, it became a theater for a short time and then Kozy Korner variety store. In 1924, it became Petchel's Café, a highly regarded place to dine en route to Duluth from the Twin Cities. In the early 1950s, it became Jimmy's, and later, it became Pat's. Since 1980, it has been Nicoll's Café. (Above, courtesy of Pine City Area History Association; below, author photograph.)

THE 61 WAY CAFÉ. The 61 Way Café, first called Pine City Café and Nu-Way, was on the corner of Highway 61 and Second Avenue SE in Pine City, seen here at far left. The popular café was

also a bus stop for Greyhound. The building still stands as a mixed-use property, but the café ceased operations in the 1990s. (Courtesy of Pine City Area History Association.)

PINE CITY MAIN STREET. Pine City's Main Street is Old Highway 61. Seen above in 1939 is a view looking south over the Snake River Bridge. On the left is the Coca-Cola Bottling Company. The view below, in 1942, looks north from Fourth Avenue. On the left are the Pine County Courthouse and the former Pine City Village Hall. (Both, courtesy of Pine City Area History Association.)

A&W All-American Food. This A&W in Pine City dates to 1955 and was built by Al Jahnke. It is notable for being one of the few remaining A&Ws with a carhop so motorists can park and order coney dogs and frosty mugs of root beer, and a server brings it out on a tray that is hung on the car window. Over the years, it has hosted Cruise the Dub classic car shows. (Author's collection.)

Rock Creek. Both Highway 61 and the railroad went right through downtown Rock Creek, but its downtown nearly vanished after Interstate 35 was constructed to the west. Rock Creek bears the name of a creek that flows south into the St. Croix River. The township of Rock Creek, organized in 1874, was incorporated as its own city in the 1970s for fear of being annexed by nearby growing Pine City. (Courtesy of Jerry Huddleston.)

Minnesota Correctional Facility. Minnesota Correctional Facility–Rush City opened in February 2000 and houses 1,000 adult males in a high-security setting. Minnesota's newest prison cost $90 million to construct, and the project stands as the largest single structure in Chisago County. The prison, just off Highway 61, employs about 350 workers, and MINNCOR Industries is housed there. (Courtesy of Minnesota Department of Corrections.)

Rush City Flour Mill. In the 1880s, Rush City was a flourishing milling town thanks to the region's large wheat crop. Its mill, once called Raddison Milling Company, changed names over the years. In 1942, it became Amber Milling until 2002, when it became Horizon Milling Co. In 2014, it became Ardent Mills until it closed in 2019. At its peak, the macaroni in three out of every four boxes of Kraft Macaroni & Cheese was made with this mill's flour. (Courtesy of Minnesota State Historical Society.)

The Grant House. Russell H. Grant, second cousin of Pres. Ulysses S. Grant, built the Grant House in the heart of Rush City in 1880 to serve travelers looking to dine or lodge during their travels. The president himself was a popular visitor, as he was often in the area trout fishing on the Sunrise River. After a fire destroyed the original Grant House in 1895, it was rebuilt. (Courtesy of Minnesota State Historic Preservation Office.)

The Raven's Nest. The Raven's Nest & Wing, referred to as just "the Raven's Nest," was constructed in 1947 on the west side of Highway 61. Known for its quality food and large steaks, it would stay open until 2:00 a.m. after dances and all night on opening day of fishing or hunting seasons. After a fire in 1962 destroyed it, the restaurant never reopened. (Courtesy of Minnesota State Historical Society.)

Kaffe Stuga. First known as Palmer's Café in the 1910s, Kaffe Stuga (pronounced "coffee stooguh") in Harris has been a cozy, roadside diner for its entire existence. The Swedish-themed eatery was first a feed mill and store before becoming a popular stop along Highway 61 for beef dinners and homemade pies. Local farmers, anglers, and avid cooks Willard and Bonnie Ramberg were the proprietors from the late 1950s through early 1960s. (Courtesy of Brian Kays.)

Merchants State Bank. Merchants Bank's downtown North Branch came about in 1895 as the second bank in the village. It changed its name to Merchants State Bank in 1904, and by 1915, it had nearly $300,000 in deposits. Community leader John Rystrom, who held various roles—North Branch mayor, Chisago County commissioner, and Minnesota senator from 1914 to 1919—was the bank's president for a time. (Courtesy of John Frid.)

North Branch Main Street. This postcard of North Branch's Main Street is from the 1910s. North Branch incorporated in 1881, and in 1901 was split between the more densely populated village and its large surrounding township. It was not until 1994 that the two merged again. North Branch is named after the north branch of the Sunrise River. (Courtesy of John Frid.)

Rustic Inn. In 1938, Gust Williams established the Rustic Inn at the main intersection in Stacy, along the east side of Old Highway 61. It has been popular with motorists and is known for its burgers. In 1949, it was sold to Stanley Olson. In 1968, it was sold to Everett and Sharon Peterson, followed by their daughter Kristy Pavek in 1999, who ran it with her son Aaron. (Courtesy of City of Stacy.)

Hallberg Marine. When Eugene "Gene" Hallberg learned about a Forest Lake boat dealership shutting down in 1958, he saw an opportunity to expand his father's car dealership right on Highway 61. It was called Hallberg Pontiac-GMC, at 620 Lake Street North. Soon, boat brands such as Alumacraft and Mercury were displayed alongside General Motors automobiles. By 1972, due to space needs, operations moved to Wyoming, and the car business was eventually sold off. The company became the largest boat dealer in the world for several brands. Today, it has one million square feet of buildings on over 100 acres of land. (Both, courtesy of Hallberg Marine.)

Three

The Southeastern Bluffs

Wyoming to La Crescent on US Highway 61

Wyoming, Minnesota, to La Crescent, Minnesota, remains the US portion of Highway 61 today. However, many of the sections of this route are marked with other names in addition to US Highway 61. For example, from the southern city limits of St. Paul to the Wisconsin state line is the Disabled American Veterans Highway. The Hiawatha Pioneer Trail runs from where 61 meets Interstate 94 in St. Paul to the Wisconsin border, and the Laura Ingalls Wilder Historic Highway is the portion of 61 from Highway 60 in Wabasha to US Highway 63 in Lake City.

Various cities along the route denote Highway 61 with a local name as well. In Wyoming and Forest Lake, for example, it is known as Forest Boulevard. In St. Paul, the route is labeled Arcade Street. In 2014, the portion of Highway 61 between Forest Lake and Wyoming was officially designated Trooper Glen Skalman Memorial Highway, commemorating the officer shot and killed in the line of duty in 1964 along that stretch of road.

This entire portion of Highway 61 in southeastern Minnesota is part of a unique geologic region called "the Driftless Area," formed by glaciers long ago. Minnesotans also call this area "Bluff Country." The entire stretch offers scenic beauty, including Barn Bluff, Sugar Loaf Bluff, and meandering roadways alongside the Mississippi River. This is part of the Great River Road scenic byway.

Historic and charming cities along this stretch of Highway 61 include Hastings, Red Wing, Lake City, Wabasha, and Winona.

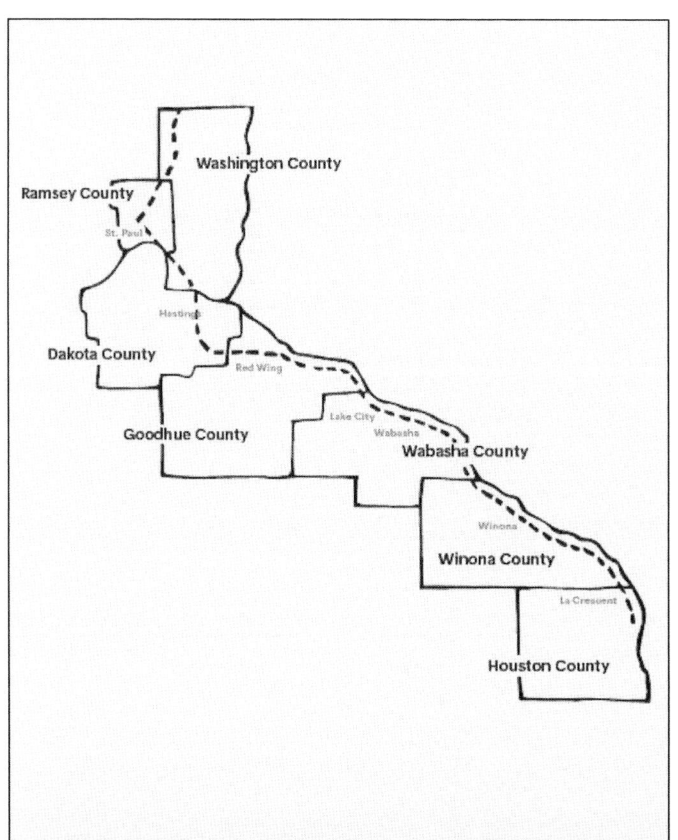

St. Paul to La Crescent. This map shows the approximate route of Highway 61 through southeastern Minnesota, largely running parallel with the Mississippi River, through Ramsey, Washington, Dakota, Goodhue, Wabasha, Winona, and Houston Counties. Here, the highway meanders along the river bluffs through charming river towns such as Hastings, Red Wing, Lake City, and Wabasha. (Courtesy of MNiCards.)

Hitching Post Motel. Forest Lake's hitching Post Motel can be seen in the 1996 Academy Award–winning movie *Fargo*. About a mile north of downtown on Highway 61 sits the motel that was reimagined in Bismarck, North Dakota, in the scene where Jerry Lundegaard (William H. Macy) was finally tracked down and arrested for the kidnapping of his wife. (Author's collection.)

FOREST THEATRE. The Forest Theatre was erected by 1916 with a seating capacity of 320, and was first called the Home Theater. It was later converted into a twin-screen theater owned by the Drummonds and Fladlands for over 70 years before it closed in May 1990 after showings of *The Hunt for Red October* and *Born on the Fourth of July*. The family also owned a drive-in theater called Hub Drive-In north of town on Highway 61. (Courtesy of Jackie Fladland.)

HOPKINS SCHOOLHOUSE. The Hopkins Schoolhouse was built in Hugo in 1928, serving as an elementary school until 1962. In 1965, it was purchased for $35,000 by Oneka Township from the Forest Lake School District for use as a town hall. It was a town hall until 1972, when Oneka Township became the Village of Hugo, later the City of Hugo. For a time, it was a youth center. (Author's collection.)

WHITE BEAR LAKE DEPOT. The White Bear Businessmen's Association is pictured in 1953 in front of the White Bear Lake depot, which was built in 1935 and still stands at Highway 61 and Fourth Street. A couple of women are present. The railroad came through town as early as 1868, connecting St. Paul and the international seaport of Duluth. It ushered in the "resort era" of the late 1800s, as travelers would come to local lakes and patronize local establishments. (Courtesy of White Bear Lake Area History Association.)

PAUL R. BEAR. The large stucco bear perched above the Ralph Thomas Chevrolet dealership was erected in 1965 at a cost of $25,000. He was given the name Paul R. Bear. Though the dealership has changed hands over the years, the 35-foot-tall, five-ton statue has remained. Today, the dealership, owned by Walser Automotive Group, is aptly named Polar Chevrolet. (Courtesy of Minnesota State Historical Society.)

NORTHERNAIRE MOTEL. The Northernaire Motel was a one-story motel built in 1955 along an old portion of Highway 61 after it was bypassed. This stretch of highway in Maplewood primarily serves traffic between White Bear Lake and St. Paul. The Northernaire had a classic neon sign for most of its existence until 2011, when it was replaced by a plastic one. (Courtesy of Jerry Huddleston.)

NORTHWESTERN HOTEL. Arcade Street is the name of Highway 61 through St. Paul, a stretch of about eight miles through the city, now interrupted. At 934 Arcade Street was once a fine hotel—Northwestern Hotel—along the highway, but over the years, it deteriorated and became a flophouse for a time. Today, it houses apartments above an award-winning bar. (Author's collection.)

Hamm's Brewery. The Theo. Hamm Brewing Co. is seen here around 1900, just blocks away and visible from Highway 61 in St. Paul. This area of the city was bustling over 100 years ago. The Hamm family mansion can be seen in the background to the right, but it burned down via arson in 1954 and was later redeveloped as parkland. (Author's collection.)

Saint Paul. This was a popular postcard sent by those visiting the capital city around the mid-20th century. Several iconic St. Paul images show through the big letters, including the Minnesota State Capitol and the Cathedral of St. Paul. (Courtesy of John Frid.)

Dakota County Courthouse. Now Hastings City Hall, the historic government building just off 61 in Hastings opened as the Dakota County Courthouse in 1871. Designed by A.M. Radcliff, the building is Italian Villa style. It was the center of government in Dakota County until 1974. In 1993, it became Hastings City Hall. (Courtesy of Douglas Kerr.)

Hastings Spiral Bridge. The Hastings Spiral Bridge's design looped wagons and horses into the downtown business district. It was constructed tall enough to let steamboats pass underneath. The roadway across the bridge was still gravel when it became US Highway 61 in 1926. It was constructed in 1895 at a cost of $39,050. It was later replaced due to wear and tear in 1951. (Author's collection.)

WIEDERHOLT'S SUPPER CLUB. Since 1932, the family-owned Wiederholt's Supper Club in Miesville has offered fine dining along this rural stretch of Highway 61. Now in its fourth generation, it was birthed from an idea George Wiederholt had in the 1920s. It grew from a grocery store, tavern, and filling station to a popular pregame stop for Mudhen baseball fans and one of the most notable supper clubs in southeastern Minnesota by the 1960s. (Courtesy of August Schwerdfeger.)

JACK RUHR FIELD. Baseball has been played in Miesville, eight miles south of Hastings, since the 1920s. The Mudhens name has been around since 1948. Jack Ruhr Field, beside Highway 61, opened in 1961, and night games began to be played in 1994 to large crowds. The field is named after Ruhr, the Mudhens' longtime president who was at the helm for all five of their state championships. (Courtesy of Jack Huddleston.)

Barn Bluff. The city of Red Wing's most famous landmark, Barn Bluff, took its current form as the result of raging glacial meltwaters. A public park since 1910, the entire bluff is considered sacred by the Dakota people because it is the site of many burial mounds. The bluff faces downtown Red Wing and is often painted with timely messages. (Courtesy of Goodhue County Historical Society.)

Red Wing Station. Red Wing Station was originally built by the Chicago, Milwaukee, St. Paul & Pacific Railroad and is now operated by Amtrak. Construction of the building began in 1904. It was designed by architect J.M. Nettenstrom in a Neoclassical Revival style. Trains still run daily from Chicago to St. Paul via Red Wing. (Courtesy of John Frid.)

Nybo's Café. Nybo's Café sat prominently on Highway 61 in downtown Red Wing for decades. In 1947, Bob Nybo Sr. installed 10 bowling lanes upstairs. In 1984, Nybo's Café and Lanes relocated to Withers Harbor Drive and expanded to 20 lanes. It was later called Nybo's Landing. In 1995, the family sold the restaurant business to the Bierstube, and in 2011, the bowling portion was sold to them as well. (Courtesy of John Frid.)

Red Wing Main Street. In the early 1850s, settlers from Mississippi River steamboats came to farm the rich land around Red Wing, encroaching on Mdewakanton Sioux territory. Red Wing lost its port prominence when flour mills were built at Minneapolis and Saint Anthony. Other major industries in town kept the city's economy robust. Main Street is US Highway 61. (Courtesy of John Frid.)

St. James Hotel. During the early part of the 1870s, the thriving wheat market in Red Wing created a need for tourists and businesspeople to have a comfortable place to stay while passing through. That resulted in this extravagant 62-room hotel along Highway 61, which still stands and is in operation today. (Courtesy of John Frid.)

Fair Trade Books. Books have been sold in Red Wing since its incorporation. But in 2014, a very unique, independent bookseller opened downtown: Fair Trade Books. The resident dog, Reveler, watches over the shop. On one's first visit, the shop owner gifts a free book to anyone who says, "Books make great gifts." More than 35,000 books have been given away since the store opened. (Courtesy of Fair Trade Books.)

HANISCH BAKERY. Hanisch Bakery began as Pirius Bakery (1944–1972) in the heart of downtown Red Wing. It was Brachler's Bakery for 35 years after that until 2007, when Bill Hanisch purchased it and it became Hanisch Bakery. The space has been a bakery since the 1920s. It is known for its specialty cakes, homemade bread, and a variety of donuts and pastries. For decades, the bakery has received extra notoriety every presidential election with its famous cookie poll that generally follows the actual popular vote. Also, the bakery's mascot, "Bunman," inspired by Bill Hanisch, can be seen around Red Wing at events and parades. (Both, courtesy of Bill Hanisch.)

Minnesota Correctional Facility–Red Wing. The Minnesota Correctional Facility–Red Wing was built in 1889 as the Minnesota State Training School. A prominent architect at the time, Warren Dunnell, designed the original Romanesque-style building. It now mainly serves as a state juvenile correctional facility, with 40 adult male prisoners in minimum-security housing. The institution was the subject of Bob Dylan's "Walls of Red Wing." (Courtesy of Goodhue County Historical Society.)

Red Wing Pottery. The land around Red Wing is rich with clay, and so what is now Red Wing Stoneware & Pottery has flourished on the north side of Red Wing since 1861. Today, a store is connected, and visitors can peek at potters through factory windows. The museum there contains a 70-gallon clay jug and holds more than 6,000 vintage pieces of stoneware. (Courtesy of Goodhue County Historical Society.)

Red Wing Shoes. Charles Beckman founded Red Wing Shoes in 1905. It grew from each of the two world wars with mass production of soldiers' boots. In 1926, the original women's boot debuted, the Gloria, with an unmistakable style and extended height. In 1952, the infamous 877 Moc Toe cap-toe style boot was created. In 1987, Red Wing Shoes acquired S.B. Foot Tanning Company, a producer of fine leathers. Although Red Wing Shoes stores can be found across America, its flagship store in Red Wing is home to the world's largest boot, which is US size 638.5. (Above, courtesy of Goodhue County Historical Society; below, courtesy of Michel Curi.)

SHELDON THEATRE. The Sheldon Theatre, Red Wing's jewel box, was built in 1904 and is listed in the National Register of Historic Places. The theater was named after Theodore B. Sheldon, a successful businessman and council member whose estate funded the project. The City of Red Wing first operated it, making it the second municipally-owned theater in the United States. It has staged many national acts. (Courtesy of Sheldon Theatre.)

FRONTENAC. Two communities comprise Frontenac, along US 61. The railway line near Frontenac Station and Old Frontenac separates residences along the bluff. In 1957, the land between and some land beyond was set aside as Frontenac State Park, which includes the Mississippi floodplain, bluffs, prairies, and hardwoods. (Courtesy of Goodhue County Historical Society.)

WATER SKIING. Ralph Samuelson, who lived from 1903–1977, was the first person to attempt what we now call water skiing. It happened on Lake Pepin by Lake City while being towed by his brother Ben, holding onto a clothesline and standing on boards. After several days and several dozen attempts, on July 2, 1922, he discovered that by leaning backward with the ski tips pointed upward, he could rise above the water, which led to his invention of the new sport. Water skiing has since been refined and has caught on around the world. Lake City celebrates Water Ski Days each year to honor the local invention. (Both, courtesy of Lake City Historical Society.)

Lake City Main Street. Lake City is at the intersection of US Highways 61 and 63 on the Mississippi River at Lake Pepin. The Sioux were the first in the area, and the town was later platted in 1855 and incorporated as Lake City in 1872. In its early years, Lake City was a port market for grain. (Courtesy of John Frid.)

Clamming. The Lake Pepin Pearl Button Company in Lake City sent button blanks harvested from Lake Pepin clams to Muscatine, Iowa, for manufacturing into buttons. The business thrived from 1914 to 1920, when clam stocks were depleted and plastic buttons became the new way. The Pearl Button Factory building still stands at 226 South Washington Street in Lake City. (Author's collection.)

NATIONAL EAGLE CENTER. The Upper Mississippi River Valley is full of bald eagles. Hundreds migrate here each winter since this stretch of river remains ice-free, allowing the eagles to fish. In 1989, Eagle Watch Inc. was formed to help the community see eagles from an observation deck. People flocked to the area to see this American icon. In 2000, the National Eagle Center opened in downtown Wabasha, and its interpretive center was upgraded in 2007. (Author's collection.)

MISSISSIPPI FLYWAY. The Mississippi Flyway is a waterfowl flyway generally following the Mississippi River and Highway 61 from Canada to the Gulf of Mexico. Good sources of food, water, and cover exist for the entirety of the flyway, but many species of birds tend to congregate in the Driftless Area along southern Minnesota portions of the Mississippi, where they are enjoyed by birdwatchers, especially on drives along 61 through Wabasha County. (Courtesy of US Fish and Wildlife Service Mountain Prairie.)

Lark Toys. Along Highway 61 in tiny Kellogg, Minnesota, is LARK Toys, with more than 20,000 square feet to explore, putting it among the largest independent specialty toy stores in the world. It houses a full-sized carousel with hand-carved fantasy animals, and includes handmade wooden toys, an 18-hole mini-golf course, and a candy shop. Toy connoisseurs appreciate Memory Lane, with toys from past eras. (Courtesy of Annie Montecillo.)

Greetings From Winona. This "Greetings From Winona" postcard was produced in 1943 by Curt Teich and depicts several recognizable Winona scenes. From left to right in the big letters are Sugar Loaf, Princess Wenonah Fountain in Windom Park, the Hot Fish Shop at the base of Sugar Loaf, the Mississippi River, Highway 61, and J.R. Watkins Company. (Courtesy of John Frid.)

The Hot Fish Shop. The Hot Fish Shop put Winona on the culinary map and was in business from 1931 to 1999. Henry and Helen Kowalewski opened it, and it was passed onto their son Lambert and his wife, Helen Kowalewski, in 1967. The third and fourth generations of the original owners became owners or operators of the establishment as well. (Courtesy of John Frid.)

J.R. Watkins Company. The Joseph Ray, or J.R., Watkins Company was founded in 1868 in Plainview, Minnesota, but moved to Winona, a thriving lumber town, in 1885. Watkins sold products door-to-door at first and grew to become the world's largest direct-sales company. A pioneer in natural living, Watkins became a household name for natural home and personal care products, such as remedies and extracts. (Courtesy of Winona County Historical Society.)

Hotel Winona, Winona, Minn.

WINONA HOTEL. The grand Winona Hotel was built in 1889. In 1983, it was listed in the National Register of Historic Places for its significant Romanesque Revival architecture. It accommodated tourists during Winona's heyday as a performing arts destination. The building is now the Kensington, and has been converted to senior apartments. (Courtesy of John Frid.)

LATSCH ISLAND. Latsch Island is named after John Latsch, a Winona benefactor. Two miles off Highway 61, it is home to a 120-plus-houseboat community with limited amenities. The island is on the Minnesota side, and its houseboats are ramshackle shanties and custom-builds. They have faced removal every decade or so, mostly during the Great Depression. (Courtesy of Visit Winona.)

Sugar Loaf. Easily seen from Highway 61, Sugar Loaf is the 85-foot-high rock pinnacle that is Winona's most distinguishing landmark. The result of 19th-century quarrying, the bluff has been part of a Native American legend that portrays it as the cap of Dakota chief Wapashsha. At the foot of the Sugar Loaf was a brewery, now repurposed, and listed in the National Register of Historic Places. (Courtesy of Winona County Historical Society.)

Bauer's Market. In 1957, John and Thecia Bauer assisted with a small apple stand along Highway 61 in La Crescent. Bauer's Market was born shortly thereafter, as well as sons who helped grow the family-run business. Since the beginning, the Bauer family's passion for apples has been on display at their store. (Courtesy of Doug Connell.)

Oaks Hotel and Night Club. Off Highway 61 in Minnesota City, six miles northwest of Winona, an 1864 hotel, saloon, and eatery became Oaks Hotel and Night Club in the 1930s. Rumored to be a mob hangout, it was also known for its food, liquor (with an 87-foot bar), and orchestra music. Put on the map during Prohibition, Lawrence Welk once played there. Its 600-capacity dining room was popular through the 1950s and 1960s. (Both, courtesy of John Frid.)

Four

The Road Trip South

Wisconsin to Louisiana on Highway 61

The United States created Highway 61 in 1928 as a north-south companion route to Route 66. It was developed from existing roads and resulted in a magnificent highway that traversed forests, farmland, rocky bluffs, floodplains, big cities, and small towns. It connected Grand Portage, Minnesota, to New Orleans, Louisiana, traversing 10 states.

Beyond Minnesota, the remainder of the 1,400-mile highway passes through Wisconsin, Iowa, Missouri, Arkansas, Mississippi, Tennessee, and Louisiana. This chapter focuses on the history and iconic parts of this notable road past the border of the Gopher State.

Highway 61 is an important part of the American story, from music along "the Blues Highway" to the advancement of the civil rights movement.

Most people know of the music element. In Minnesota, there's Bob Dylan, of course, but also alternative rock legends Bob Mould and the Replacements. Then, there are the many famous jazz and blues artists whose story began on the southern stretches of the highway.

Fewer know this highway also played an important role as a route for many African Americans to escape the harsh realities of Jim Crow. Displaced sharecroppers traveled north on 61 out of the Delta to Memphis and beyond. Today, the National Civil Rights Museum in Memphis explores the history of the civil rights movement and its impact on the culture today.

It all ties together, as the five million or so who set out northbound brought the blues with them. The southern Highway 61 towns like Tunica, Clarksdale, Vicksburg, and Natchez were left with storied histories.

Highway 61 Through the United States. This map shows the 1,400-mile route of Highway 61 from the Canadian border to New Orleans, passing through Minnesota, Wisconsin, Iowa, Missouri, Arkansas, Tennessee, Mississippi, and Louisiana. The route largely follows the Mississippi River and goes through major cities along the way, including St. Paul, St. Louis, Memphis, and New Orleans. (Courtesy of MNiCards.)

Joseph B. Funke Candy Company. Now the boutique Charmant Hotel, this was once the largest of three candy operations in La Crosse, Wisconsin. It ran in the early 1900s, and in 1920, Funke made 160-plus chocolate varieties in addition to 500-some other candies. The factory closed in 1933 during the Great Depression, later becoming Ross Furniture. (Courtesy of Charmant Hotel.)

World's Largest Six-Pack. La Crosse has long been known for its love of adult beverages, having among the nation's highest number of bars per capita. In 1969, a brewery along Highway 61 made its six enormous beer tanks resemble a six-pack of Old Style and La Crosse beer cans. Across the street from the six-pack is a replica of Gambrinus, the "King of Beer." (Courtesy of Tony Webster.)

Digger's Sting. Digger's Sting has occupied the former La Crosse Chop House building since 1985, when local restaurant and lounge owner Victor Skaff opened the swanky restaurant that specializes in succulent prime rib. The restaurant was named after the 1973 film *The Sting* and a New York mafia hitman, "Digger." The gangster era is still alive here today. (Courtesy of Robby Virus.)

POTOSI BREWING COMPANY. Beer was brewed in Potosi, Wisconsin, since 1852. It was once the fifth-largest brewery in Wisconsin and was shipped coast to coast. It survived Prohibition but closed in 1972. In 1980, it was listed in the National Register of Historic Places. In 2004, it was chosen as the site of the National Brewery Museum. There is still an active microbrewery on the premises today. (Courtesy of Wisconsin Historical Society.)

DICKEYVILLE GROTTO. In the heart of Dickeyville, right along Highway 61, stands a grotto covered in a plethora of gems, glass, shells, and other brightly colored objects donated by area churchgoers. Nearby Holy Ghost Parish's priest Mathius Wernerus built it in 1925 as a memorial to three local boys killed in World War I. The grotto is dedicated to religion and patriotism. (Courtesy of Kelly Ludwig.)

Fenelon Place Elevator. The Fenelon Place Elevator is a funicular rail line in Dubuque, Iowa. Billed as the world's shortest, steepest railway, it runs 296 feet along one of the town's many limestone bluffs. Wealthy businessman J.K. Graves built it to ease his commute from his home atop the hill to the bank where he worked near the bottom. Modernized in the 1970s, it remains open to the public for a nominal fee. (Courtesy of *Telegraph Herald*.)

Dubuque Shot Tower. This Dubuque landmark reaches 150 feet into the sky and is one of America's last remaining shot towers. Built in 1856 to provide lead shot for the military, it never operated to its capacity, in large part due to the Great Panic of 1857. It was also used as a fire watchtower for a period, and today is listed in the National Register of Historic Places. (Courtesy of Annie Montecillo.)

Hotel Julien. At Dubuque's Hotel Julien, named for 18th-century voyageur Julien Dubuque, elegance has always been apparent. The hotel is in the Old Main District and may be considered Iowa's most famous hotel. Still in operation, it is rumored to have accommodated Abraham Lincoln on his travels to Illinois and to have been once owned by mob leader Al Capone. (Courtesy of Hotel Julien.)

National Mississippi River Museum. Dubuque's waterfront was given new life in 2003 with the addition of the National Mississippi River Museum, dedicated to the history and culture of "Old Muddy." The museum complex doubled in size in 2010 and is also home to an aquarium, casino, and the steamboat *William M. Black*, a national landmark. Garrison Keillor narrates the museum's introductory film *River of Dreams*. (Courtesy of Richie Diesterheft.)

The 61 Drive In Theatre. Five miles south of Maquoketa, Iowa, amidst farmland and just west of Delmar, is the 61 Drive In Theatre. It has been there since 1950, with a capacity of 180 cars. Twice—in 1960 and again in 1995—screens were replaced due to storms, more recently a tornado. It was the first drive-in in the Midwest to switch to digital projection and FM stereo sound. (Courtesy of 61 Drive In Theatre.)

The Filling Station. The Filling Station was established in 1971 along the Highway 61 Business Route in Davenport, Iowa. Resembling an old filling station, the restaurant and bar was filled with gas station memorabilia. Proprietor Donny Wachal, who started and ran the business for five decades, died in 2020 due to COVID-19-related complications. (Courtesy of the Filling Station.)

Von Maur. In 1872, an immigrant and his three sons opened a small storefront in downtown Davenport. Over time, it moved and expanded into its flagship store at the intersection of Main and Second Streets in the Redstone Building. After mergers, including with the retailing family the von Maurs, it later became the first Von Maur store. The Redstone Building today is home to the River Music Experience, a multi-use music facility. (Courtesy of Davenport Public Library.)

Municipal Stadium. Davenport's riverfront ballpark was called Municipal Stadium from the time it was built in 1931 until 1971, when it was renamed John O'Donnell Stadium until 2007. Today, it is Modern Woodmen Park, where the Quad Cities River Bandits play home games. A cornfield is planted down the left-field line, and players run out from it before games, as a nod to the 1989 film *Field of Dreams*. (Courtesy of Davenport Public Library.)

National Pearl Button Museum. Pearls were sent from Midwestern waters from points north, such as the Highway 61 towns of Pine City and Lake City, Minnesota, to Muscatine, Iowa. Muscatine was the undisputed "Pearl Button Capital of the World" when the trade hit its peak between 1908 and the 1920s. The Muscatine History & Industry Center became a national museum dedicated to this industry in 2019. (Courtesy of Freshwater and Marine Image Bank at the University of Washington.)

Maid-Rite. In 1926, someone tasted butcher Fred Angell's loose meat sandwich with a selected blend of spices and said, "This sandwich is made right." With that, Maid-Rite was born with a single store in Muscatine, and it soon became a chain. Four of the original locations are still in operation today, the second home to the first carhop in America. There are over 30 Maid-Rite locations in five states. (Courtesy of Mike Willis.)

Snake Alley. Snake Alley is a crooked street built in 1894 in Burlington, Iowa. In 2017, Ripley's Believe It or Not! Recognized the street as the Crookedest Street in the World and the No. 1 Odd Spot in its Odd Spots Across America campaign. Originally an extension of North Sixth Street, it was intended to link downtown with the Heritage Hill Neighborhood. (Courtesy of coalfather.)

Main Street Bed & Breakfast. At the northwest corner of Main and Center Streets in Hannibal, Missouri, a building was constructed in 1876 that housed Farmers & Merchant's Bank. Today, the upper portion is home to genteel lodgings in the midst of Hannibal's historic downtown district. (Courtesy of Michael Ginsberg.)

Garth Woodside Mansion. Dating to the 19th century, the Garth Woodside Mansion in Hannibal was owned by John and Helen Kercheval Garth, close friends of Samuel Langhorne Clemens—better known as Mark Twain. Over the years, the pair entertained Twain at their home numerous times. Letters about these visits remain. It is thought that John Garth inspired the character of Tom Sawyer. Today, late-19th-century antiques adorn the guest rooms and cottages. (Courtesy of Jamie Ricks.)

Mark Twain's Boyhood Home. Writer Mark Twain's boyhood home, a museum since 1912, is in the 200 block of Hill Street in Hannibal. Twain lived in the grey house from 1844 to 1853, and inspiration for many of his stories, including his best-known work, *The Adventures of Tom Sawyer*, came while living here. The museum hosts annual events and is home to historical collections. (Courtesy of Library of Congress.)

The Champ Clark House. This national historic landmark in Bowling Green, Missouri, was once home to Speaker of the US House of Representatives Champ Clark. He was also a presidential candidate in the 1912 primaries that Woodrow Wilson emerged from. A Kentucky transplant, Clark moved to Missouri in 1898 to open a law practice. He nicknamed the estate "Honey Shuck." (Courtesy of Liz Rosenburg.)

Bankhead Candies. Bankhead Candies on Highway 61 has been producing hand-dipped candies since October 1919. The candies are prepared in large copper kettles, poured onto a marble table, and then cut. Thomas Jefferson Bankhead, a great-grandson of Pres. Thomas Jefferson and a prominent Bowling Green restaurateur, began the operation. It started as a restaurant with a side candy business, but now candy is the main attraction. (Courtesy of Bankhead Candies.)

ROADHOUSE 61. The Highway 61 Roadhouse in Webster Groves, Missouri, northwest of St. Louis, was established in 2006. It celebrates all the great foods of the Blues Highway region from St. Louis to New Orleans. On the menu are Cajun, Creole, Midwestern comfort food, barbecue, and a combination of the latter—barbecue spaghetti, a Memphis staple. Inside, blues music plays. (Courtesy of Crystal Rolfe.)

THE CHESHIRE. The Cheshire in St. Louis has a history dating to the 1920s, when it was a hamburger stand. The Medarts expanded it to add Olde Cheshire, a tavern. Over the years, it became a restaurant and hotel, and in 2011, it went through a multimillion-dollar renovation to restore it as a boutique hotel designed to feel like a traditional British inn. (Courtesy of the Cheshire.)

IMO'S PIZZA. Ed and Margie Imo founded Imo's Pizza in 1964 near the Hill neighborhood of St. Louis. They made square-cut pies with Provel cheese in place of Mozzarella, a distinct taste key to their growth and success over the years. Ed and Margie's children now run Imo's, and they have grown it to over 100 stores and franchises in the Greater St. Louis area, as well as to places beyond such as Illinois and Kansas. (Courtesy of Brian Johnson and Dane Kantner.)

CATHEDRAL BASILICA OF ST. LOUIS. Ground breaking on the Cathedral Basilica in the Central West End part of St. Louis took place in 1907, but the super-sized Catholic cathedral did not have its first Mass until 1914, when it was finally complete. It is one of two basilicas in St. Louis; the other, Basilica of St. Louis King of France, is near the iconic arch. (Courtesy of Jonathan Cutrer)

The Gateway Arch. The world's tallest arch dominates the skyline of St. Louis. It was designed by architect Eero Saarinen in 1948 and was constructed between 1963 and 1965. Officially called Jefferson National Expansion Memorial, the stainless steel structure reaches 630 feet into the air, the same distance between its two legs. It represents a doorway to westward expansion. (Courtesy of Josh Hallatt.)

Forest Park. West of downtown St. Louis, near Washington University, is Forest Park, 1,300 landscaped acres known as the "Heart of St. Louis." It opened in 1876, more than a decade after it was planned. It hosted the 1904 World's Fair (the Louisiana Purchase Exposition) and the 1904 Summer Olympics. Today, it includes the St. Louis Zoo, art and history museums, and a science center. (Courtesy of Christian Collins.)

SAINTE GENEVIEVE. Originally a French Canadian town, Sainte Genevieve is the oldest town in Missouri, and its historical aspects are its major draw. Pictured here is the old post office. Houses date back 200 years, some of which have been preserved, while some have been left ramshackle. The town was named after the patron saint of Paris. (Courtesy of Library of Congress.)

SOUTHERN HOTEL. One of the oldest hotels west of the Mississippi River is the Southern Hotel, which towered over the downtown of thriving Sainte Genevieve. The Federal-style building was erected in the 1790s, but it has operated as a hotel since 1805 and was widely known by river travelers as the finest accommodations between Natchez and St. Louis. (Courtesy of Paul Sableman.)

Kaskaskia Bell. Fifteen miles south of Sainte Genevieve, Kaskaskia, Illinois, is a virtual ghost town now, thanks to numerous floods, but it was once the capital of Illinois. Kaskaskia held on as the only Illinois settlement west of the Mississippi River. It is home to the 650-pound Kaskaskia Bell, also known as "the Liberty Bell of the West," a gift from France. (Courtesy of Kbh3rd.)

Cape Girardeau. Cape Girardeau—or as residents call it, "Cape"—was occupied by Union forces during the Civil War. Despite fires and flooding, much of the vibrant historic business district remains intact from before the Civil War. It comprises one of the largest historic districts in the country. Within it is the Glenn House, built in 1883 for merchant and banker David Glenn. Today, it is a museum. (Courtesy of Paul Sableman.)

Lambert's Café. In 1942, Earl and Agnes Lambert borrowed $1,500 to start Lambert's in a small building on Main Street in Sikeston, Missouri. World War II and rationing made it difficult, but the family-run business hung on. Known for its pass-around side dishes and hot rolls being thrown to customers, business began to flourish. Tens of thousands of cans of sorghum molasses are used each year to top the "throwed rolls," hundreds of dozens of which are made daily. By 1981, Lambert's again outgrew its space, relocating and expanding. Constant growth meant another move in 1988 to the highway. Two additional locations have since opened in Ozark, Missouri, and Foley, Alabama. (Both, author's collection.)

HUNTER-DAWSON MANSION. The Hunter-Dawson mansion in New Madrid was built in 1859 for William and Amanda Hunter, local storeowners. Typical of pre–Civil War Mississippi River plantations, it now serves as a historic site. There are 15 rooms and nine fireplaces, and much of the original furniture is still on display. Daughter Ella and her husband, William Dawson, inherited the home before donating it to the state. (Courtesy of Missouri State Parks.)

US HIGHWAY 61 ARCH. At the border of Missouri and Arkansas, north of the city of Blytheville, is an arch over the highway that was built in 1924. When the highway was paved that year, the Mississippi County Road Improvement District built it. It was listed in the National Register of Historic Places in 2001. (Courtesy of Annie Montecillo.)

BASSET POW CAMP. Just north of the small town of Bassett, Arkansas, visible from Highway 61, was a World War II prisoner of war camp built to house about 300 German and Italian prisoners. The POWs picked cotton for less than a dollar a day. It was closed in 1946, and little evidence remains of the site, save for the gate at the entrance and this historical marker. (Author's collection.)

HU. HOTEL. Hugh Lawson White Brinkley, or "Hu" as he was called, made his fortune and put Memphis on the map as a shipping hub. He was a grandson to one of the original founders of Memphis. He also planned the building that's now home to Hu. Hotel. Ahead of his time, he also enjoyed yoga, lived modestly, and worked to improve women's health in honor of his wife. (Courtesy of Annie Montecillo.)

PEABODY HOTEL AND DUCK PARADE. Though it was built in 1869 and hosted guests like Andrew Johnson and William McKinley, the elegant Peabody Hotel became a cornerstone of the downtown Memphis area after a grand reopening ceremony in 1981. Today, it is most famous for its feathered residents, the Peabody Ducks, who live on the roof and make twice-daily treks to the lobby to swim in the magnificent hotel fountain—a tradition since the 1930s that has become an internationally famous attraction. The duck parade started as a joke among a couple of friends. Bellman Edward Pembroke, a former circus animal trainer, served as the Peabody Duckmaster for 50 years until he retired in 1991. (Both, author's collection.)

GRACELAND. Graceland was once part of a 500-acre family farm named after one of the family members, Grace. The mansion was built in 1939. Eighteen years later, when rising star Elvis Presley was just 22, he purchased the estate for a little more than $100,000. Though he would own homes elsewhere, this was always home base for Elvis. (Courtesy of Library of Congress.)

LORRAINE MOTEL AND NATIONAL CIVIL RIGHTS MUSEUM. The National Civil Rights Museum is attached to the restored Lorraine Motel in Memphis, where Dr. Martin Luther King Jr. was assassinated in 1968. The museum, which has been open since 1991, underwent a major renovation in 2014. It follows the struggle for civil rights and the culture of resistance dating to the first slaves brought to America in 1619. (Courtesy of Brian Rawson-Ketchum.)

ROCK 'N' SOUL MUSEUM. Memphis's Rock 'n' Soul Museum on legendary Beale Street was created by the Smithsonian Institution in 2000 and tells the story of musical pioneers who overcame racial and socioeconomic barriers to create music that would have a profound impact on the world. It was the first permanent Smithsonian exhibition outside of New York or Washington. (Courtesy of Ron Cogswell.)

MEMPHIS PYRAMID. At 321 feet tall, the Memphis Pyramid was first known as the Great American Pyramid; after all, it was the sixth-largest pyramid on earth. It was originally constructed in 1991 to serve as a 20,000-plus-seat basketball arena, and despite going dark for a period, since 2015 it has been a flagship store for Bass Pro Shops. (Courtesy of Tyler Merbler.)

BEALE STREET ENTERTAINMENT DISTRICT. One of America's most iconic streets, Beale Street in Memphis is three blocks of restaurants, shops, and nightclubs that play live delta blues, jazz, rock 'n' roll, and R&B. A. Schwab, established in 1876, is the oldest store on Beale and the last remaining original business. It was not until the early 1900s that Beale became honky-town central after musician William Christian Handy moved to town. (Courtesy of Annie Montecillo.)

PAULA RAIFORD'S DISCO. Established during the height of disco dancing in 1976, Robert Raiford opened Raiford's in a small building a few blocks south of downtown Memphis. It had a successful run until it closed in 2007. After that, he and his daughter Paula opened Paula Raiford's Disco, a two-story nightclub downtown. It is open today as one of the last homes to disco in America. (Courtesy of Sean Davis.)

ARCADE RESTAURANT. The Arcade is Memphis's oldest restaurant. Speros Zepatos founded it in 1919 after emigrating from Greece; it was the Arcade Diner until 1925. Located on South Main Street a couple blocks off Highway 61, it was the hub of thriving Memphis in the mid-1960s. The King of Rock 'n' Roll, Elvis Presley, was a regular here. There was a period of decline through the 1970s, but it is thriving again today. (Courtesy of Sean Davis.)

JERRY'S SNO CONES. About 100 years ago, Jerry's was a Sinclair fueling station. It is believed the owner made snow cones for kids whose parents had their vehicles worked on. New owners L.B. and Cordia Clifton revived the tradition in 1967 as a car wash and snow-cone stand. It became known for its Snow Cone Supreme with ice cream. (Courtesy of Jerry's Sno Cones.)

CHARLIE VERGOS' RENDEZVOUS. Charlie Vergos' Rendezvous is a c. 1948 barbecue spot in a Memphis basement, serving up racks of smoked ribs and sausages with dry-rub spices. Originally, it served as the basement of his diner, but Vergos discovered a coal chute that he made into a barbecue pit. He later converted his diner into the famous restaurant it is today and moved the entrance from the street to the back alley. (Courtesy of Alexandra Kay.)

SUN RECORDS COMPANY. Sun Records is a recording label in Memphis founded by producer Sam Phillips in 1952. It was where Elvis Presley, Roy Orbison, and Johnny Cash recorded their historic 1950s tracks. Before then, Sun mainly recorded African Americans because of Phillips's love for R&B and his desire to bring it to a caucasian audience. Ike Turner recorded the very first rock 'n' roll song in the Sun studio, "Rocket 88." (Courtesy of Library of Congress.)

Tunica Resorts Casinos. Formerly Robinsonville, the Tunica Resorts is a destination about 30 miles south of Memphis built mostly in the 1990s, after lawmakers made casino gambling legal for Mississippi counties next to the river. The first casino, Splash, opened in 1992, followed by other big names in the casino world. It became the third-largest American gambling center, but following the 2011 floods and competition, only six casinos remain. (Both, courtesy of Sean Davis.)

GATEWAY TO THE BLUES. The Gateway to the Blues Museum and Visitor Center is in a renovated train depot just north of Tunica, Mississippi. An iconic neon sign signals the center, where dozens of guitars, historical artifacts, hands-on exhibits, and a small recording studio highlight the blues. The entire place is a launchpad for exploring the Delta. (Courtesy of Library of Congress.)

BLUE & WHITE RESTAURANT. On 61, further down the Blues Highway in Tunica, is the iconic Blue & White Restaurant, serving hearty meals and southern hospitality since 1924. It has been in its current location—a former Pure Oil fueling station—since 1937. Not much has changed since it opened. The menu includes fresh donuts in the morning, fried green tomatoes, and later, flash-fried pickles. (Courtesy of Blue & White Restaurant.)

Devil's Crossroads. The Devil's Crossroads, or "the Crossroads," is a trio of electric guitars on a pole at the intersection of old US Highways 49 and 61 in Clarksdale, Mississippi. It marks where legend says in the 1930s, musician Robert Johnson sold his soul to the devil in return for the ability to master the blues. Prior to this, Johnson was said to be a good harmonica player but terrible on the guitar. (Courtesy of Library of Congress.)

Shack Up Inn. Once a working plantation, Hopson Plantation, just south of Clarksdale, the grounds of the Shack Up Inn still include a cotton gin and other outbuildings and equipment. The plantation sat dormant from 1972 to 1995, when it became a southern-style bed and breakfast with authentic sharecropper shacks that visitors can stay in. Today, there are over 40 rooms and an onsite Chapel Bar. (Courtesy of Visit Mississippi.)

DELTA BLUES MUSEUM. The Delta Blues Museum has existed since 1979 in an effort to preserve, interpret, and encourage interest in the story of the blues. The museum has since relocated to the historic Clarksdale Passenger Depot in the heart of Clarksdale's "Blues Alley" district downtown. The museum houses artifacts such as the boyhood shack of blues legend Muddy Waters. The depot is listed in the National Register of Historic Places. (Courtesy of Visit Mississippi.)

RIVERSIDE HOTEL. Riverside Hotel in Clarksdale has been in operation since 1944. Before that, it was the G.T. Hospital, where renowned blues singer Bessie Smith died after her 1937 car accident. It became a well-known hotel, one of the few in Mississippi that provided lodging for African Americans at the time, hosting such musicians as Duke Ellington, Ike Turner, Robert Nighthawk, and Sonny Boy Williamson II. (Courtesy of Visit Mississippi.)

Ground Zero Blues Club. Ground Zero is a blues club that opened in Clarksdale in 2001, co-owned by actor, director, and narrator Morgan Freeman. The name was fitting since Clarkdale is known as "Ground Zero" for the blues. It was opened in a repurposed grocer and cotton grading company that sat vacant for decades. Local musicians and touring acts perform on its stage, to the delight of blues fans. (Courtesy of Annie Montecillo.)

Birthplace of Kermit the Frog Museum. Jim Henson is America's most famed visionary and puppeteer. A Muppet museum in the small Mississippi town of Leland honors the roots of Henson and his most beloved creation, Kermit the Frog. Henson grew up and played among the area's swamplands full of frogs. Henson died in 1990, but his Muppets live on. (Courtesy of Annie Montecillo.)

HIGHWAY 61 BLUES MUSEUM. In the former Montgomery Hotel building in Leland, some major influences in blues and rock 'n' roll music are highlighted in this small museum. Nearly 150 musicians, some world-famous, most from mid-Mississippi, have their histories preserved here. (Courtesy of Annie Montecillo.)

THE DUFF GREEN MANSION. The Duff Green Mansion in Vicksburg, Mississippi, just off 61, has served many purposes over time. It was built in 1856 for local cotton broker Duff Green and his bride-to-be. During the Civil War, Green allowed his home to be used by both Confederates and Union troops as a hospital. Later, it served as a boys' orphanage, headquarters for the Salvation Army, and today, as a lavish bed and breakfast. (Courtesy of Library of Congress.)

BIEDENHARN'S COCA-COLA MUSEUM. The Biedenharn Candy Company building houses a Coca-Cola Museum in downtown Vicksburg. It features the equipment that Joseph Biedenharn used in 1894 to bottle Coke for the first time. Today, the museum has exhibits about the beginnings of Coca-Cola, the Biedenharn family, and the process used to first bottle the soda. Coca-Cola advertising and memorabilia are also on display. (Courtesy of Annie Montecillo.)

VICKSBURG NATIONAL MILITARY PARK. Vicksburg is home to the Vicksburg National Cemetery and National Military Park. More than 17,000 Union soldiers (three-quarters of whom are unknown) from the Battle of Vicksburg are buried in the cemetery. The events of the early 1860s and the reasons why Vicksburg was vital to victory are commemorated in the park. (Courtesy of National Park Service.)

Old Country Store. Established in 1875, the Old Country Store in the unincorporated community of Lorman, Mississippi, still stands. It has had a lot of past lives: post office, Western Union office, bus depot, train station, telephone office, craft mall, ballroom, Louisiana-style restaurant, and

a bank. Since 1996, Arthur Davis's restaurant has served his famous fried chicken and a buffet of southern comfort foods like catfish, cornbread, ribs, and turnip greens. (Author's collection.)

BIRTHPLACE OF THE TEDDY BEAR. Onward, Mississippi, is the birthplace of the Teddy Bear. A historical plaque marks the place where, in 1902, Mississippi governor Longino led Pres. Theodore "Teddy" Roosevelt on a bear-hunting excursion. Supposedly, the president refused to shoot the defenseless bear, considering it unsportsmanlike. The *Washington Post* ran cartoons about the incident; from them, inventor Morris Michtom made the bear smaller and cuter and, later, produced them among his other toys. (Courtesy of Tim Evanson.)

CHOCTAW HALL. Prior to the Civil War, Natchez had the most millionaires per capita in America. That led to a plethora of antebellum homes being constructed, many of which remain today and serve as a testament to the history of the South. One of them is Choctaw Hall, built around 1836. Its four stories and 11,600 square feet feature a blend of styles from Greek Revival to Federal. (Courtesy of Choctaw Hall.)

RHYTHM NIGHTCLUB. On St. Catherine Street in Natchez, a memorial commemorates a packed nightclub set ablaze in 1940 in what is believed to be an accident that killed over 200 people and injured many others. The event is recounted in several songs, including Howlin' Wolf's 1956 "The Natchez Burning." Since 2010, blues fans from around the world have been able to learn more about the fire at the Rhythm Club Museum. (Courtesy of Visit Natchez.)

MUSEUM OF AFRICAN AMERICAN HISTORY AND CULTURE. Opened in 1991 in a 1904 post office building in Natchez, the Museum of African American History and Culture chronicles black heritage in the southern United States. The 10,000-square-foot space showcases exhibits and events from 1716 to the present, covering slavery, the Civil War, Reconstruction, 20th-century wars, and the civil rights era. (Author's collection.)

Mammy's Cupboard. Though scorned by some, one of the most iconic roadside restaurants along Highway 61 is Mammy's Cupboard, just six miles south of Natchez. The restaurant and gift shop are essentially a 28-foot-tall sculpture with horseshoe earrings. First painted in 1940 to look like a black woman, it was repainted after the 1960s civil rights movement to downplay its racist overtones. (Courtesy of Library of Congress.)

Wilkinson County Courthouse. There is an extraordinary courthouse in Woodville, Mississippi, the county seat of Wilkinson County, just 10 miles north of the Louisiana border. It was a rare Beaux-Arts design by J.R. Gordon, constructed in 1903–1904. The interior rotunda has two balconies and a tile floor inlaid with the county seal. The courthouse anchors the town square of Woodville. In 1973, a two-story expansion was added, which included a jail. (Courtesy of Keith Vincent, courthousehistory.com.)

ROSEMONT PLANTATION. About a mile east of Woodville is Rosemont Plantation, the boyhood home of the Confederacy's president, Jefferson Davis. It was built in 1810 with wooden pegs holding together salvaged posts and beams, a makeshift way to construct the Jefferson family's cotton plantation. The property is home to a large rose garden planted by Davis's mother, for which the plantation was named. (Author's collection.)

BELLUE'S FINE CAJUN CUISINE. A few miles north of downtown Baton Rouge along Highway 61 is Bellue's Fine Cajun Cuisine, owned by Wirt Bellue. The restaurant and butcher shop has been serving Cajun dishes, such as boudin (rice and meat) and gumbo pots since the late 1960s. The Cajun influence is strong in Baton Rouge, while Creole is a little stronger in New Orleans, to the south. (Author's collection.)

Bourbon Street. Tulane Avenue, or Highway 61 through New Orleans, takes one a few blocks from Bourbon Street. Calle de Borbón was the name of the street when New Orleans was the capital of the Spanish Province of Louisiana from 1762 to 1803. Today, it is in the festive French Quarter of the city and is the pulse of New Orleans tourism, full of bars, restaurants, and entertainment. (Courtesy of Corey Balazowich.)

Café Du Monde. The original Café du Monde was set up in New Orleans's French Market in 1862. Its name translates to "Café of the World" in French. It has been located in the same vicinity ever since, with its current building erected in the 1930s. The coffee shop is famous for its beignets (French-style donuts lavishly covered in powdered sugar) and its chicory-blended coffee (developed by Creoles), which adds a chocolate-like flavor. (Courtesy of Annie Montecillo.)

HOTEL MONTELEONE. Perhaps the most famous hotel in Louisiana, the Beaux Arts–style Hotel Monteleone, built in 1886, is in the historic French Quarter, the only high-rise nearby. It was beloved by authors such as Eudora Welty and Ernest Hemingway. It has 600 guest rooms and the famed Carousel piano bar, complete with a 25-seat revolving bar. (Courtesy of Brian Lauer.)

The two men in this photograph are unidentified. They pose with both a US Highway 61 marker and one of the original state trunk route markers on the same pole. These two markers were only present from the fall of 1926 through the spring of 1934, when the US highways were just an add-on on top of the state trunk routes. Minnesota incorporated them into the numbering system in 1934. (Courtesy of Andrew Munsch.)

Index

Bibliography

Mayo, William, and Kathryn Mayo. *61 Gems on Highway 61: Your Guide to Minnesota's North Shore: From Well-known Attractions to Best-kept Secrets*. Cambridge, MN: Adventure Publications, 2018.

Lange, Jessica. *Highway 61*. Brooklyn, NY: PowerHouse Books, 2019.

Bright, Derek. *Highway 61: Crossroads on the Blues Highway*. Choir Press, 2020.

McNally, Dennis. *On Highway 61: Music, Race, and the Evolution of Cultural Freedom*. Berkeley, CA: Counterpoint, 2015.

Wurzer, Cathy. *Tales of the Road: Highway 61*. St. Paul, MN: Minnesota Historical Society Press, 2008.

About the Organizations

Carlton County Historical Society
www.carltoncountyhistory.org

Chisago County Historical Society
www.chisagocountyhistory.org

Cook County Historical Society
www.cookcountyhistory.org

Dakota County Historical Society
www.dakotahistory.org

Goodhue County Historical Society
www.goodhuecountyhistory.org

Houston County Historical Society
www.houstoncountyhistoricalsociety.org

Lake County Historical Society
www.lakecountyhistoricalsociety.org

Old Highway 61 Committee
www.oldhighway61.com

Pine City Area Historical Association
www.pinecityhistory.org

Pine County Historical Society
www.pinecountyhistoricalsociety.org

Ramsey County Historical Society
www.rchs.com

St. Louis County Historical Society
www.thehistorypeople.org

Wabasha County Historical Society
www.wabashacountyhistory.org

Washington County Historical Society
www.wchsmn.org

Winona County Historical Society
www.winonahistory.org